DISCARD

WEST GA REG LIB SYS
Neva Lomason
Memorial Library

THE INCREDIBLE WORLD OF PLANTS

THE GREAT PLAINS

CHELSEA HOUSE PUBLISHERS
New York • Philadelphia

Text: Andreu Llamas
Illustrations: Luis Rizo

Las Grandes Llanuras © Copyright EDICIONES ESTE, S. A.,
1995, Barcelona, Spain.

The Great Plains copyright © 1996 by Chelsea House
Publishers, a division of Main Line Book Co. All rights
reserved.

1 3 5 7 9 8 6 4 2

Library of Congress Cataloging-in-Publication Data

Llamas, Andreu.
  [Grandes llanuras. English]
  The great plains / text, Andreu Llamas : illustrations, Luis
Rizo.
      p. cm. — (The Incredible world of plants)
  Includes index.
  Summary: Describes the grassland ecology of the great plains
of the world with their various landscapes such as the savannah,
the tundra, the steppe, and the great American plains.
  ISBN 0-7910-3464-X. — ISBN 0-7910-3470-4 (pbk.)
  1. Grassland ecology—Juvenile literature. 2. Grassland
plants—Juvenile literature. [1. Grasslands. 2. Grassland ecology.
3. Ecology. 4. Plants.] I. Rizo, Luis, ill. II. Title. III. Series: Llamas,
Andreu. Incredible world of plants.
QH541.5.P7L5813   1995
574.5'2643—dc20   95-2947
                                                    CIP
                                                    AC

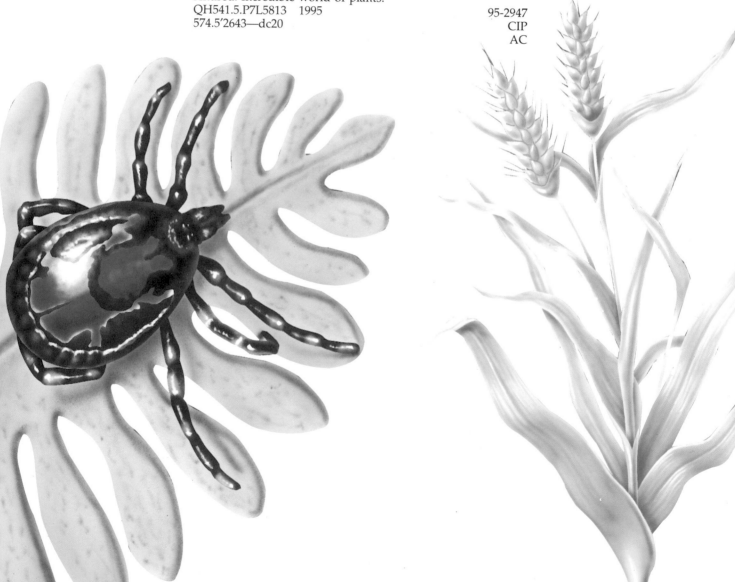

CONTENTS

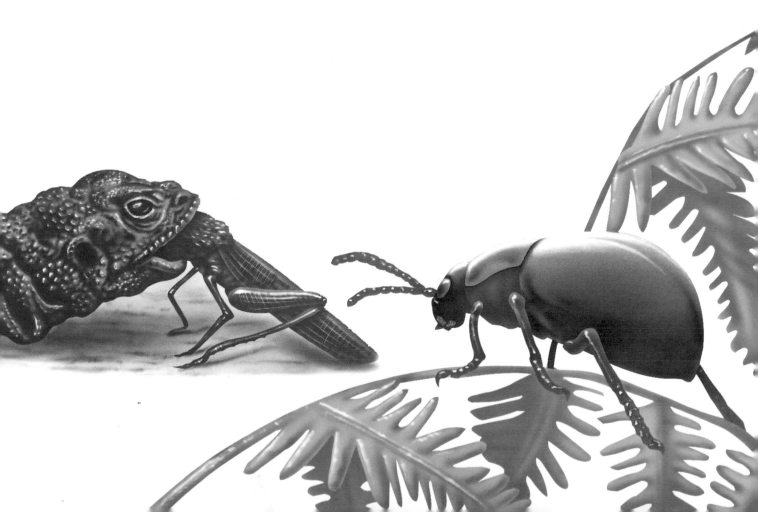

THE AFRICAN SAVANNA

Have you ever heard of the African savanna?

The savannas are huge wide open spaces with a few shrubs and trees spread out here and there. The trees are not very big, but they have a powerful system of roots that takes the utmost advantage of the soil's moisture. Some of these trees, such as some acacias, have flat treetops like that of a parasol. Other trees, like the baobab, have very thick trunks that are capable of storing large quantities of water.

The ground is covered by a dense layer of grass that can reach up to 7 feet (2 meters) in height. The herbaceous plants dry up and disappear for the most part during the dry season. This is when these grasses form a ''blanket'' of yellowish and extremely flammable straw, which is subject to periodic fires. There are many different kinds of savannas: the wet savanna, the thorny savanna, and the dry savanna. In some areas the trees are mostly palms and make up what are called palm savannas. The different savannas also have different appearances. There are some savannas that are always green, some with trees that lose their leaves in the summer, and others with trees that lose their leaves in the cold season.

In the driest areas there are even cacti and a few isolated patches of grass separated from each other by a large expanse of bare earth. Grass, however, is the protagonist of the savanna. For example, elephant grass forms almost impenetrable masses which grow up to 16 feet (5 meters) tall.

(1) The acacias
The acacias are the most representative trees of the African savanna. Their roots collect enough water in the dry periods. They also have thorns to defend themselves from being attacked or eaten.

(2) A bright reptile
The agama lizard is one of the easiest reptiles to spot. Only the adult males capable of beating their rivals keep their bright blue, red, and orange coloring.

(3) Serpents in the savana
This vibrant nocturnal serpent is one of the many that live in the savana. But do not worry—its venom is not dangerous for humans.

(4) Very efficient trash collectors
Not even 15 minutes after an elephant defecates, more than 3,800 dung beetles are capable of coming together to ''attack'' and dig through the feces. In a half-hour there is hardly any left.

3

THE GRASSES OF THE SAVANNA

Did you know that grass is the best-suited plant to resist periodic droughts, fires, and the constant presence of herbivores?

It is difficult to imagine how plants can withstand the constant feeding of thousands of the savanna's herbivores especially since the grass provides food for a number of the world's large mammals. The most abundant kind of grasses is the gramineous variety, which includes hundreds of different edible species.

Unlike trees and bushes, grass grows from the base of the stem instead of the tip, allowing it to grow again after having been burned by fire or eaten by herbivores.

Grasses do not require insects for their reproduction because they rely on the wind or reproduce themselves through underground shoots.

The blanket of grass on the great plains is the result of the struggle and the simultaneous evolution of both the grasses and the herbivores. Grass "defends" itself with its fast growth and regeneration process but it also uses other techniques: hardened tissues, thorns, and the accumulation of toxic or repellent substances. Some grasses can even intermingle with other nonedible plants and hide.

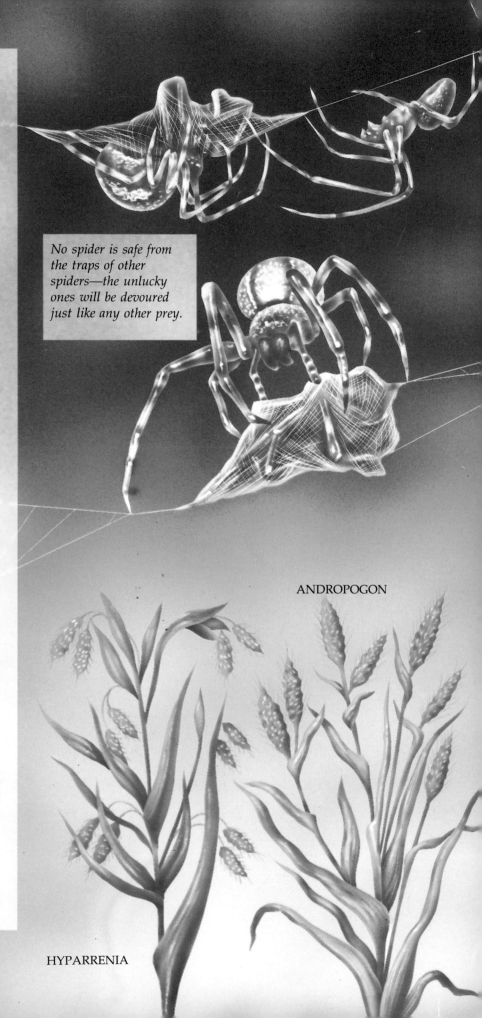

No spider is safe from the traps of other spiders—the unlucky ones will be devoured just like any other prey.

ANDROPOGON

HYPARRENIA

Even though they may look alike, the many varieties of gramineous plants are very different from one another. Here are some of the most common.

PANICUM MAXIMUM
OR GUINEA GRASS

PENNISETUM PURPUREUM
OR ELEPHANT GRASS

FIRE IN THE SAVANNA

Fire plays a major role in the makeup of the savanna, given that fires occur there every year. As the dry season advances, the greens that once covered the savanna slowly lose their intensity.

In just a short time, the dried up remains of most grasses completely cover the great plains while the tree branches lose all their leaves. As you can imagine, the savanna's appearance during the drought is yellow, dry, dusty, and even looks dead. This is when even the smallest spark can cause a fire.

At the beginning of the dry season, grass still retains enough water, and fire only destroys the first layer of leaves and dead grass. This is known as low, or cold fire, which does not harm the trees. This type of fire only raises the temperature a 1/2 inch below the ground so as a result, the roots and seeds are not burned. Animals are also safe because they can take refuge in their dens.

However, as the dry season advances, so does the number of fires that become much more destructive and extensive. Did you know that many trees are all the same age precisely because of the fires? This is because regrowth occurs when fire does not affect an area for a period of seven or eight years.

These periodic fires are responsible for the grasses dominating the savannas because they do not kill the underground parts of the herbaceous layer but they do limit the number of trees to only a few of the fire resistant species.

(1) Fire resistant
The savanna's trees are resistant to the fires that affect the lower layer during the dry season.

(2) No defense
For insects, small mice, and other small animals, fire is a frightening enemy that provokes catastrophes within their populations.

(3) Destroyed trees
Younger trees and sprouts are very vulnerable to fire. Ninety percent of the plants under 3 feet (1 meter) tall and seventy percent of the plants under 7 feet (2 meters) tall are destroyed. However, tall trees are hardly affected.

(4) Escape by flying
Fire does not have negative consequences for the animals who can escape the flames due to their wings or large size.

2

RAIN IN THE SAVANNA

The savanna also has flowers. During the wet season, flowers appear in the prairie with their bright colors.

The seasons in the savanna begin and end with this flowering. It is prompted by the monsoons, which are the winds that the Indian Ocean blows in from the northeast for six months each year and then blows in from the southeast bringing the much-awaited rainfall. In general, the big rains occur in the spring, while in November and December come the small rains. Did you know that between November and June there is enough moisture in the savanna to allow the grass to grow in only one day? But during the rest of the year the ground dries up and the grass grows slower and thinner.

The rhythm of life on the savanna is heavily influenced by the rains. For example, you have probably heard about the great migration of the *ungulates* when the dry season occurs. The rainy season ends little by little. Near the end of spring, the green of the grass begins to gradually change to yellow. At this point, many of the herbivores seem to realize that the drought is coming and they get quite nervous. It is time to begin the trek once again in search of greener pastures.

(1) Annual rains
During the humid season, very violent rains beat down on the great plains. The dry land finally receives the necessary water.

(2) The terrifying ant lion
The ant lion lives in soft and sandy soil, where it builds a trap in the shape of a funnel. Small insects, like ants, fall down the small slope and are trapped by the jaws of the ant lion. The ant lion may even throw grains of sand at its prey to make them fall into its trap.

(3) Reptiles of the savanna
Many reptiles live in the savanna, like this gecko lizard with its thick toes. To escape its enemies it detaches its tail.

(4) Apartments for birds
The republican birds build communal nests. Over 100 pairs can live together. The nests are so heavy that they can break the tree that supports them.

2

CROPS INSIDE THE DARKENED CASTLE

Did you know that termite societies are the most incredible, numerous, and complex of all insects?

There are more than 2,000 species of termites that live in nests of many different shapes and sizes. These nests can be enormous, measuring more than 20 feet (6 meters) in height and 98 feet (30 meters) in diameter at the base. Other nests, in contrast, are underground and house only a couple hundred termites. Termites have an incredible capacity for controlling the temperature inside their nest.

Some nests have more than 5 million termites inside. These nests are so complex that they even have "air conditioning"—there is a hole in the ceiling of the main chamber that the termites can widen or reduce by adding or taking away dirt particles. This allows them to modify the speed at which hot and humid air rises and escapes from the nest.

In some areas, elephants pull out trees, and the great abundance of dead wood helps the termites. The termites' role in the savanna is very important: by eating dry wood they quickly transform it into fertilizer, which is easily assimilated by the ground.

There are some termites that tend mushroom gardens. They use plants and wood that they carry to their nest to prepare the soil, rich in cellulose, for cultivation. Here they grow the mushrooms capable of transforming the lignin. These underground gardens are shaped like wood balls on which the mushrooms grow. These gardens are dark in color, spongy looking, and can be as large as a human head. Actually, they are an accumulation of tiny spheres made from chewed wood.

(1) Climate control
Inside the nests there is a completely different climate to that on the outside. Light never enters. The air is always calm and the temperature is constant 24 hours a day, the whole year around. Also, the amount of moisture inside is always maintained very high, even if that means that the termites have to dig more than 33 feet (10 meters) underground in search of water.

(2) Hunter ants
Termites are not the only social insects in the savanna. The hunter ants, for example, live in underground nests that house a few hundred individuals. When they go out to hunt, very few of the smaller sized animals can defend themselves. This includes the termites.

(3) Passing along nutrients
This worker is passing along food to a termite soldier. This "mouth to mouth" sharing of food strengthens the ties among the different members of the colony.

(4) Cultivators of mushrooms
Here you can see how the nest of the African termite, which grows mushrooms, is constructed. A short time after a new colony is started, the workers make the first wood balls (4a). They take pieces of vegetable matter and patiently grind them in their jaws until they get a spherical pasty mass (4b). Afterward, all the workers join their spheres together and form a single ball, growing in size with every new one added (4c).

1

2

4a

4b

4c

5

13

THE ACACIAS

One of the most typical inhabitants of the savanna is the acacia. There are 38 different species of acacias that make up 45 percent of the total arboreal vegetation.

The acacias can grow in the most diverse African environments. One of the most common acacias is the A. drepanolobium, because it is fire resistant. The taller A. clavigera is more abundant on the slopes, while A. tortilis with its typical umbrella top is found on the edges of the prairie.

In order to survive in such a difficult environment, the acacias have had to make several adaptations. For example, the trunks of acacias that are over three years old are fire resistant. When external conditions are dry, the acacias grow stunted and take on the appearance of a shrub, growing very close to the ground. It later holds up its flat and spread-out treetop, which reminds us of a beach umbrella.

Some other acacias protect themselves against the herbivores using thorns and barbs. But they cannot chase away all herbivores with their thorns. For example, a giraffe's mouth is covered with a thick corneous layer that permits it to eat, despite the acacia's thorns.

Among the many acacias, the whistling acacia stands out. It gets its name from the sounds the wind makes when blowing through the numerous prickles it uses to protect itself.

Take careful note of the weaver bird's delicately structured nest. They are a marvel of animal architecture.

The fly catcher's nest has a false and attention-grabbing entrance, while the real entrance is hidden. This throws off track the predators who come too close.

The acacia forms gallnuts *in reaction to different insect bites, and ants install themselves inside these. In exchange, the ants chase away other insects and many herbivores.*

The umbrella shape of the acacia is unmistakable. The weaver birds decorate the acacia's branches with their basket-shaped nests. Almost every branch is occupied by one of these baskets.

THE STEPPES

What do we mean by great plains?

The great plains include very different landscapes such as the savanna, the tundra, the steppe, and the great American plains.

The steppe and the savanna are intermediate zones between forests and deserts. They take up ample regions of the continents' interior and have a dominant herbaceous layer with a few isolated shrubs and trees.

The steppe can be found in zones with a very hot summer and a cold winter. The steppe sometimes receives the name of low grass prairie because it is made up of smaller sized scattered grasses. Isolated shrubs and small trees can also be found on the steppe, but the amount of ground cover is scarce and a large quantity of bare earth is exposed.

In the steppe there are many varieties of gramineous grasses. One gramineous plant common in the American steppe is buffalo grass. Other typical plants are the sunflower and the poisonous plant, the locoweed.

The tangled and dispersed esparto grass or mat-weed copses are the most common environments on the steppe.

There are many different kinds of steppes. The high barren steppes are very hard, solitary, and inhospitable, but the steppe does not always look so hostile. In the spring, the landscape is enriched with a multitude of yearly plants also known as pteridophytes.

The shepherd's purse has adapted to the harsh conditions of the plains by adopting a spherical shape and pointed stems with no leaves.

The agama lizard of the steppe, which lives in Asia, follows the shadows that the insects make as they jump. In doing so, the agama lizard can find and capture the insects.

These beautiful butterflies (the milkweed) have large wings that allow them to glide for several miles with ease.

The conditions on the steppe during the winter are very difficult with strong frosts. Here you can see a thistle covered by a cold morning's frost.

The earwig uses its pincers to feed on the animal, vegetable, fruit, and small insect remains it finds. These pincers or claws that seem so frightening are in reality harmless, so do not be afraid when you see one.

THORNY PLAINS

On the African continent there are enormous expanses of extremely dry areas known as the thorn kingdom.

Acacias of different species and sizes, from the dwarf acacias (which look like bushes) to the giant acacias (which can grow more than 66 feet high with a flat treetop more than 98 feet in diameter), can be found here. Should you want to go for a walk through this shrubby steppe you would soon realize how difficult it is to go even a few steps without being pricked by the many barbs of the dwarf acacias. You would also be able to notice that under the hot sun, the bark on the shrubs glistens with metallic flashes, wearing out your eyesight and disorienting you.

In the shrubby steppe, the scarcity of water has made the plants evolve and perfect different strategies for obtaining and storing water. The acacias, for example, have powerful roots that penetrate deep within the earth, extending themselves radially to find as much water as possible. Their trunks are also thick and waterproof. This way they have an effective barrier against evaporation.

On the other hand, their tiny leaves only appear a few days before the rainy season. Afterward, the leaves grow increasingly stunted as the dry season returns.

The gramineous grasses that live in the shrubby steppe have also adapted themselves to the drought, and when the rains come they grow rapidly, producing seeds that are highly resistant to dehydration. These seeds later sleep in the sandy soil for several months, waiting for the next rains so that they can begin to *germinate*.

The hornbill transforms the hollow of a dry tree into a fortress. While the female is still inside, her mate makes the entrance smaller with hardened mud. The female then stays walled up with her chicks for the entire nesting period while the male feeds them from the outside.

When the honey-seeker, or pointer, finds a wild beehive, it leads other animals to the entrance with jumps and trills. It later takes advantage of what is left. They are good partners. One discovers the beehive, and the other destroys it.

An infinite number of birds live in the shrubby steppe. Among the branches of the thorny shrubs there are dozens of small birds perched, like these jaybirds that feed on the grasses.

This saurodactyl has such a thin and delicate skin that if you try to pick it up, you will be left holding only bloody pieces. But it can regenerate its skin in only two or three days.

When the rains arrive, the grays of the shrubs are slowly covered up by tiny dark green leaves, which are also very nutritious.

THE GIGANTIC BAOBAB

The enormous baobabs are the largest trees on the planet and without a doubt some of the oldest. You should know that some of these specimens are more than 2,000 years old and 30 feet (9 meters) thick in diameter!

For this reason, among all the trees that cover large portions of Africa, the gigantic baobabs always stand out above the rest. These giants normally grow far from each other or in small groups.

Sometimes the baobab is called the backwards tree because on top of its impressive barrel-shaped trunk, there are only a few thin branches that look like a handful of roots.

As you know, the acacia and the thorny thicket are specialized in terms of saving water. The baobab, on the other hand, has perfected its storage. The secret behind this success lies in its ability to store water in its tissues. Its trunk is like a gigantic barrel whose tissues can store many gallons of water during the long and much-feared droughts.

When the first rains come, the "naked" baobabs sprout green leaves, and a little later large flowers some 5 inches (12 centimeters) across in diameter. In the late afternoon the flowers open, and they are *pollinated* by bats and magpies. They have to hurry, though, because the flowers drop off within 24 hours.

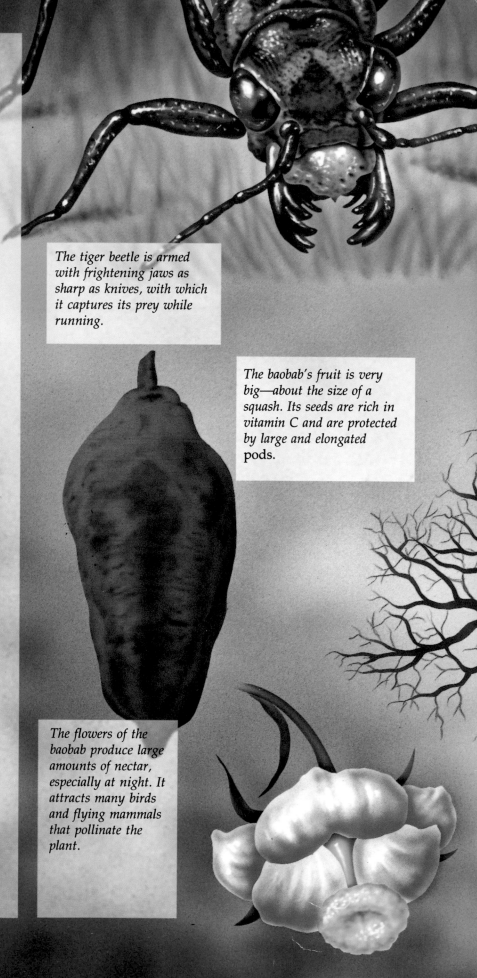

The tiger beetle is armed with frightening jaws as sharp as knives, with which it captures its prey while running.

The baobab's fruit is very big—about the size of a squash. Its seeds are rich in vitamin C and are protected by large and elongated pods.

The flowers of the baobab produce large amounts of nectar, especially at night. It attracts many birds and flying mammals that pollinate the plant.

The harlequin lizard has a sharp ear, allowing it to detect its enemies, especially snakes, as they approach.

The baobab is unmistakable. Surely it does not look like any other tree that you can think of.

THE GREAT AMERICAN PRAIRIES

The great American prairie is located in the middle of the American continent, far away from oceans. It is a veritable "sea of grass" almost 13,000 feet long and 3,300 feet wide.

The prairies are made up of almost every type of herbaceous vegetation. Trees and shrubs are almost completely absent, although in this same region you can find fragments of forests. The great prairies have a very special climate. There is a strong contrast between the extremely hot summers and the cold winters. Another of its characteristics is the tremendous violence of the storms and hurricanes that reach this open territory. The prairies were full of immense herds of bison, pronghorns, and prairie dog colonies, as well as millions of migratory birds.

The grass has deep roots and forms a continuous and dense turf that flowers in the spring and at the beginning of the summer. The prairie's landscape is dominated by the gramineous grasses, but the leguminous and composite grasses are also represented.

There are three types of prairies: tall, medium, and short grass. The tall grass prairies are the richest and most complex of all, with grasses growing from 5 to 6 feet high. They are well protected against erosion because of the thick vegetation. Today, this type of prairie is considered one of the world's most important granaries because the crops grown here produce fantastic harvests.

(1) Millions of bisons
The bison used to range over most of the American prairies and reached 75 million in number, but today there are only tens of thousands left.

(2) Fighting and exhibiting
The roosters of the American prairie developed a complicated ritual to attract females. Between fighting against others and exhibiting, the best would attract and attain the females.

(3) Hiding under foam
The larvae of the foam cicada expels a soapy looking liquid through their mouths with which they form a layer to cover their entire body, using it as camouflage against would-be predators.

(4) A traveling butterfly
The monarch butterfly undertakes an incredible journey of some 3,000 miles (5,000 kilometers) in one year. It is at the end of fall that they head toward the warmer countries in the South, and then in the spring they head back North.

(5) The prairie dog
One member of the colony is always standing on a small rise keeping guard and ready to sound the alarm if necessary. Prairie dogs live in colonies, which are like huge cities housing millions.

(6) Insect births
The plant louse does not hatch from an egg, but rather, the female gives birth to a completely formed newborn.

5

THE SOUTH AMERICAN PAMPA

Although Africa is covered by more savannas than other continents, there are similar landscapes in South America, India, and Australia.

In South America, between the Chaco and Patagonia regions, the immense and monotonous plains of the pampa can be found. The pampa was formed by the continuous sedimentation caused by water and winds. The pampa's landscape is impressive, given that its fundamental characteristic is the almost complete absence of trees on these plains that extend themselves well beyond the horizon (over 193,000 square miles).

In the beginning, this enormous expanse of land was covered only by herbaceous plants, especially the gramineous grasses, thus forming an infinite sea of grass. Civilization started changing the makeup of this region little by little by planting trees and crops and by establishing grazing lands. However, there are still huge expanses of untouched land left.

The pampa's animals are different from those of other areas of the world. Among these, the running birds like the nandu (which is over 3 feet tall) and the capybara stand out. The capybara is the largest rodent in existence sometimes measuring more than 3 feet (1 meter) and weighing over 110 pounds (50 kilos)!

The cattle or sheep tick waits in thickets for an animal to pass by. It then jumps on the animal, grabbing it with its claws and suckers. It then makes a hole in the skin to suck out the animal's blood. It can suck for several days while its body swells more and more.

This is what the red-legged wading bird or saria looks like. Farmers domesticate them to protect their farm birds from snakes.

The baker bird's nest is very special because it is made from clay and is shaped like an old bread oven but only 1 foot (30 centimeters) in diameter.

The shrubs of the pampa make a good hiding place for insects, reptiles, and small mammals.

The few trees in the pampa offer excellent vantage points, so many birds use them to detect their prey.

FROZEN PLAINS: THE TUNDRA

The tundra is located in areas of the world between 70 degrees *latitude* and the snow and permanent ice-covered zones.

Here the high temperature in summer generally does not get above 50 degrees Fahrenheit (10 degrees Celcius). This is why the lower layer of the soil is always frozen. Roots can only grow a few inches in many instances!

It can be said that in the tundra there are only two seasons: summer and winter. During the long winter, darkness rules almost completely for nine months while during the summer, the sun shines permanently over the horizon for three months. During the long summer days, the outer layer of ice that permeates the ground melts. However, the ground frozen 3 inches (8 centimeters) below the surface does not let the water escape, and so it helps to form large marshes or *bogs*.

In the summer, the tundra is a flat landscape full of lakes where only a few resistant vegetal species grow. Most of the vegetation is made up of moss and lichen, which can endure the extreme temperatures. There are also a few shrubs and small trees. The plants of the Arctic herbaceous tundra are small and mostly herbaceous because the snow and the dry winds of winter destroy any plant sticking above the snow. Also, the size of the plant is controlled by the roots breaking during the freeze and the outer layer's thawing. The flooded plain provides excellent conditions for the reproduction of aquatic birds that can undertake flights of over 3,000 miles to breed here during the short summer.

(1) Life awakens
When the thaw arrives, the yearly plants begin to grow so that the tundra starts to look like an immense field spotted with puddles. In a few days, millions of insects emerge from the eggs laying in the puddles. There is activity everywhere. All the animals are in a hurry to regain strength and mate before the frosts return.

(2) A great hunter
The winter owl is accustomed to hunting both during the tundra's long winter night as well as the one continuous day that lasts three months in the summer.

(3) The fleawort
The fleawort, with its bright yellow flowers, is one of the plants that grows in the mossy marshes.

(4) Lichen
The tundra is dominated by lichen, and they alone are responsible for the sustenance of immense herds of ungulates, caribou, and reindeer. Here you can see the orange lichen that covers the tundra's rocks with its bright colors.

(5) A poisonous beetle
The blood beetle is poisonous. To warn its prey it begins to bleed from its mouth.

(6) Colorful flowers
In the spring, the bell-flowers decorate the tundra with their colors. They survive the winter because the snow keeps them covered.

(7) Mosquitoes
The large expanse of tranquil water is an ideal place for the mosquitoes' reproduction. Here you can see how the adult, recently formed, is about to leave the water for the first time.

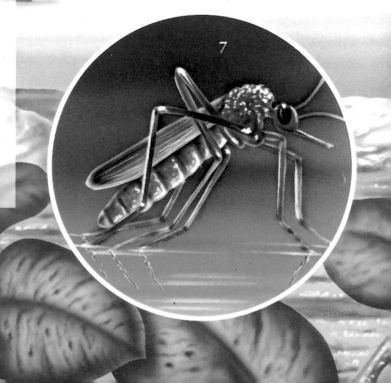

7

THE LEMMING INVASION

One of nature's longest held mysteries is that of the lemmings.

The lemming is the most typical rodent of the tundra. Its small body is very well prepared for the cold. Its wide and hairy hands and feet make it easy to travel over the snow. In the winter it takes refuge in deep tunnels that are almost 66 feet long, protecting itself from the low temperatures outside that can be 62 degrees Fahrenheit below zero. The temperature inside the den can be about 32 degrees Fahrenheit.

When the conditions are favorable, all the lemming newborns survive, resulting in a population explosion because of all the *litters*. When this occurs the number of dens can increase from 1 to 60 per *acre*! So many hungry lemmings destroy 80 percent of the vegetation in just a few days. When the lemmings can no longer find food, they migrate in search of sustenance. This happens every four years.

This lemming migration is an incredible sight. The lemmings advance, invading the paths and fields, devouring all the plants as they go. They cross water and every possible barrier. When the lemmings eventually reach the ocean, they throw themselves in without thinking twice and they sink under the waves, drowning due to exhaustion.

Previously it was thought that the lemmings were committing suicide but now some scientists believe that the lemmings throw themselves into the ocean thinking that it is just a big lake that they have to cross.

The result is that the surviving lemming population returns in just a few days to its previous numbers. A new cycle then begins.

(1) Spectacular reproduction

All animal hunters take advantage of the lemmings' migration. A female lemming with two litters per year can have up to 40 newborns when they all survive.

(2) An insect stowaway

The meloid is pretty, but it possesses powerful poison. After hatching from its egg, the larva climbs into a flower and waits for a bee to arrive. It then grabs onto the bee's hairs and lets itself be taken back to the beehive. Once inside, it feeds on the nectar and pollen that the bees have stored there.

(3) An armored wasp

The wasps from the chrysidid family lay their eggs in other bees' and wasps' nests. When discovered by the nests' owners, they roll into a ball. The owner does not recognize the wasp and throws it out of the nest. Once outside, the wasp opens its wings again and waits until the nest's owner leaves so it can go back inside and lay its eggs. The bee will then feed the wasp's eggs without realizing that they are not hers.

(4) Perfect camouflage

The white lily beetle larva camouflages itself by piling up its own feces on its back.

4

THE CROPS

In many of the planet's regions, mankind has been the main creator of great plains. For centuries people have cut down forests, changed the course of rivers, and drained marshes, all to obtain more land for crops.

Some of the first crops in human history were cereals such as millet, oats, and barley. The extensive cultivation of these crops, which are some of the most primary and the simplest, have been compared with the steppes.

Some species of animals have managed to perfectly adapt themselves to take advantage of the humans' crops. The starling, the corvidae, and the sparrow can all severely damage the crops in Europe.

In Africa, some even more destructive birds live. They are the cheleas and are greatly feared by farmers. They are typical inhabitants of the savanna, and they take advantage of the grasslands as if they were a grain and seed storehouse. They also use the isolated trees of the savanna to nest. These birds can also endure long periods of water shortage, and they are able fliers, too. The accessibility of these crops has helped increase the bird's population spectacularly, with flocks numbering one million birds! The damage they cause is comparable to that of a swarm of locusts.

(1) Free food
On the numerous acres where crops grow, animals that occupy the land take advantage of the food that is available in great quantities.

(2) Crop eaters
For African agriculture, this band of birds that live in the savanna is a true nightmare. They totally destroy the crops that grow on the land.

(3) A "tactful" hunter
The ladybug larva has very bad eyesight, and so it feels around plant stems with its legs until it finds a plant louse colony and begins to devour them.

(4) A boastful butterfly
The sesiidae is a harmless butterfly that imitates the wasp's shape, type of flight, and coloring so that nothing attacks it.

(5) An essential job
The common worm spends its life digging tunnels. Thanks to its work, the soil is aerated and oxygenated, and the different classes of soils are mixed. This way the soil becomes more fertile and productive. Earthworm concentrations can reach enormous numbers.

(6) The crab spider or tomisid
Its camouflage is perfect. It can wait several hours without moving at all before launching its surprise attack!

6

GLOSSARY

acre a square area of land measuring 4,840 square yards; equal to .405 hectare

bogs swampy land with soft mud deposited on the bottom

defecate to expel excrements

gallnuts bumps that appear in some trees due to certain insects biting and laying their eggs inside the tissues of plants

germinate when a seed begins to develop

latitude the angular distance from any point on the planet to the equator

litter the group of babies born at the same time from certain female animals such as dogs and lemmings

pod a covering or sheath in which some plants store their seeds

pollinate the transport of pollen from the anthers (the male part of the flower) to the stigmas (the female part)

ungulates a group of mammals (such as horses), the majority of which are herbivores, whose last digits of the hand or foot are completely covered by a nail or hoof

INDEX

The great plains /
J 574.52643 LLAM

763129

Llamas, Andreu.
WEST GEORGIA REGIONAL LIBRARY